SPACE FRONTIERS

Space Science

Helen Whittaker

A+

This edition first published in 2011 in the United States of America by Smart Apple Media.

Smart Apple Media
P.O. Box 3263
Mankato, MN, 56002

First published in 2010 by
MACMILLAN EDUCATION AUSTRALIA PTY LTD
15–19 Claremont Street, South Yarra 3141

Visit our website at www.macmillan.com.au or go directly to www.macmillanlibrary.com.au

Associated companies and representatives throughout the world.

Library of Congress Cataloging-in-Publication Data

Whittaker, Helen, 1965-
    Space science / Helen Whittaker.
        p. cm. — (Space frontiers)
    Includes index.
    ISBN 978-1-59920-574-8 (lib. bdg.)
    1. Space sciences—Juvenile literature. I. Title.
    QB500.22.W48 2011
500.5—dc22
                        2009038480

Edited by Laura Jeanne Gobal
Text and cover design by Cristina Neri, Canary Graphic Design
Page layout by Cristina Neri, Canary Graphic Design
Photo research by Brendan and Debbie Gallagher
Illustrations by Alan Laver, except page 28 by Richard Morden

Manufactured in China by Macmillan Production (Asia) Ltd.
Kwun Tong, Kowloon, Hong Kong
Supplier Code: CP December 2009

Acknowledgments
The author and the publisher are grateful to the following for permission to reproduce copyright material:

Front cover photos of giant radio telescopes in the Very Large Array, New Mexico © Jonathan Larsen/Shutterstock; blue nebula background © sololos/iStockphoto.

Photographs courtesy of:
© Steve Cole/Getty Images, **10** (bottom); Universal/The Kobal Collection, **30**; ESA/EADS Astrium, **8**; ESA, NASA and Felix Mirabel (French Atomic Energy Commission and Institute for Astronomy and Space Physics/Conicet of Argentina), **27**; IAU, **12** (constellation diagrams); NASA/CXC/J. Forbrich(CfA), **21**; NASA, ESA and the Hubble Heritage Team (STScI/AURA) – ESA/Hubble Collaboration, **3, 20**, back cover; NASA, ESA, The Hubble Key Project Team and The High-Z Supernova Search Team, **18** (left); NASA, ESA, M. J. Jee and H. Ford (Johns Hopkins University), **26**; NASA/ESA/R. Sankrit and W. Blair (Johns Hopkins University), **9** (all); NASA/Johns Hopkins University Applied Physics Laboratory/Southwest Research Institute/Goddard Space Flight Center, **23**; NASA/JPL-Caltech/Potsdam Univ, **6–7**; NASA/JPL-Caltech/S. Willner (Harvard-Smithsonian Center for Astrophysics), **5**; NASA/JSC, Eugene Cernan, **4**;Barnaby Norris, **13**; Photolibrary/Victor De Schwanberg/SPL, **29**; © 3d brained/Shutterstock, **10** (compass); © jamalludin/Shutterstock, **10** (binoculars); © Dmitry Karlov/Shutterstock, **10** (torch); SOHO (ESA & NASA), **19** (top); U.S. Air Force/Senior Airman Joshua Strang, **24**.

Images used in design and background on each page © prokhorov/iStockphoto, Soubrette/iStockphoto

CONTENTS

Glossary Words

When a word is printed in **bold**, you can look up its meaning in the Glossary on page 31.

SPACE FRONTIERS

A frontier is an area that is only just starting to be discovered. Humans have now explored almost the entire planet, so there are very few frontiers left on Earth. However, there is another frontier for us to explore and it is bigger than we can possibly imagine—space.

Where Is Space?

Space begins where Earth's **atmosphere** ends. The atmosphere thins out gradually, so there is no clear boundary marking where space begins. However, most scientists define space as beginning at an altitude of 62 miles (100 km). Space extends to the very edge of the universe. Scientists do not know where the universe ends, so no one knows how big space is.

Exploring Space

Humans began exploring space just by looking at the night sky. The invention of the telescope in the 1600s and improvements in its design have allowed us to see more of the universe. Since the 1950s, there has been another way to explore space—spaceflight. Through spaceflight, humans have **orbited** Earth, visited the Moon, and sent space probes, or small unmanned spacecraft, to explore our **solar system**.

Spaceflight is one way of exploring the frontier of space. Astronaut Harrison Schmitt collects Moon rocks during the Apollo 17 mission in December 1972.

SPACE SCIENCE

The study of the universe is called space science. Scientists study space phenomena, or natural events, and objects in space such as stars and planets. They also develop theories about space.

Observational Space Science

Scientists observe space objects and phenomena with telescopes. These telescopes detect sources of **electromagnetic radiation** in the universe. Electromagnetic radiation can range from long radio waves to very short gamma rays. Visible light, microwaves, and X-rays are other types of electromagnetic radiation.

▼ This image of a **galaxy** combines data gathered by the *Hubble Space Telescope* (visible light), the *Spitzer Space Telescope* (infrared radiation), and the *Galaxy Evolution Explorer* (ultraviolet radiation).

Did You Know?

The study of space is also called astronomy. The word astronomy comes from the Greek words *astron*, meaning star, and *nomos*, meaning law.

Theoretical Space Science

Scientists also develop theories to explain space phenomena. They suggest how objects in space would behave if the theory were correct and look for data that will either support the theory or prove it wrong. They also revise existing theories to take into account new observational data.

A TIMELINE OF SPACE SCIENCE

This timeline shows just a few important developments and discoveries in the history of space science.

250 BC	0	AD 1500	AD 1600

280 BC Around this time, the Greek astronomer Aristarchus suggests that the Sun, not Earth, is at the center of the solar system. Almost no one takes him seriously.

130 BC Another Greek astronomer, Hipparchus, develops the first accurate star catalogue, an astronomical list of more than 850 of the brightest stars, around this date.

AD 1543 A book by Polish astronomer Nicolaus Copernicus is published. It suggests the Sun is at the center of the solar system. People begin to take his views seriously.

AD 1608 Dutch–German lensmaker Hans Lippershe invents the telescope.

AD 1609 Italian astronomer and mathematician Galileo Gali improves on Lippershey's telescope design.

AD 1669 British physicis and mathematician Isaac Newton makes the first reflecting telescope, which used one curved mirror to reflect light and create an image.

AD **1750s** French astronomer Nicolas Louis de Lacaille goes to South Africa to observe the southern sky and begins to compile a catalogue of more than 10,000 stars.

AD **1781** British–German astronomer William Herschel discovers the planet Uranus.

AD **1846** German astronomer Johann Galle discovers the planet Neptune.

AD **1916** German-born physicist Albert Einstein publishes his general theory of relativity, which explains that **gravity** has an effect on the shape of space and the flow of time.

AD **1916** German astronomer Karl Schwarzschild proposes the idea of **black holes**.

AD **1925** American astronomer Edwin Hubble proves that galaxies exist outside the Milky Way.

AD **1931** Belgian priest Georges Lemaître proposes the **big bang theory** of the origin of the universe.

AD **1937** American Grote Reber, an amateur astronomer, builds the first radio telescope, which detects radio waves rather than visible light.

1972 AD American astronomer Charles Thomas Bolton proves that black holes exist.

1990 AD The *Hubble Space Telescope*, named after Edwin Hubble, is launched.

AD **2002** An international team of astronomers finds evidence that a source of radio waves at the center of our galaxy, known as Sagittarius A*, is a giant black hole.

2003 New data indicates that the universe is 13.7 billion years old.

2005 Astronomers find the first of several distant objects orbiting the Sun that are larger than Pluto.

2006 Pluto is reclassified as a **dwarf planet**.

PROFESSIONAL SPACE SCIENCE

Scientists have a range of sophisticated equipment to help them study space. They can see farther and study the universe in more detail than ever before.

Different Ways of Observing Space

Scientists observe most space phenomena and objects by using telescopes which detect the electromagnetic radiation they give off. Different types of telescopes detect radiation from different areas of the **electromagnetic spectrum**.

▼ Technicians inspect the mirror of the *Herschel Space Observatory* during assembly. *Herschel*, launched in May 2009, detects radiation in the infrared area of the electromagnetic spectrum.

Observing from Earth

Earth's atmosphere blocks out many types of electromagnetic radiation, but it lets through visible light and radio waves. Scientists use ground-based optical telescopes, which gather and focus visible light to observe planets, stars, galaxies, and **nebulae**. Radio telescopes allow astronomers to study things that cannot be detected in visible light, but can be detected in radio waves. These include **pulsars**, **quasars**, **cosmic background radiation**, and the remains of **supernovae**.

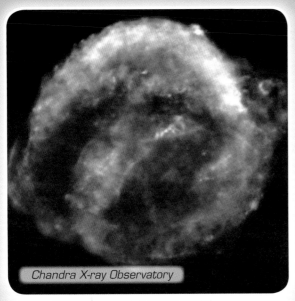

Chandra X-ray Observatory

Hubble Space Telescope

Did You Know?

Optical observatories are usually located on mountains to take advantage of the thin, clean air and to escape light pollution from cities. Radio observatories are often situated in valleys to shield them from Earth-based sources of radio waves.

Observing from Space

Space observatories are positioned in orbit above Earth's atmosphere. Some detect visible light, others detect infrared radiation, X-rays, or gamma rays. Infrared telescopes detect heat. X-ray and gamma ray telescopes detect high-energy emissions from sources such as **neutron stars**, supernovae, and black holes.

Spitzer Space Telescope

◄ These images show the remnants, or debris, of Kepler's Supernova as viewed by three space-based telescopes detecting (from top) X-rays, visible light, and infrared radiation.

BACKYARD ASTRONOMY

You do not need a telescope or other expensive equipment to study the night sky from your own backyard. You can become an amateur astronomer with just a few everyday items.

Tips for Backyard Sky Watching

Sky watching is best done on a night when there is little cloud cover. Give your eyes 10 minutes or more to get used to the dark. Then, use a compass and star charts to identify the star patterns you see in the sky and record your observations.

▼ Many households have a pair of binoculars, making it easy for anyone to view the night sky. However, some amateur astronomers prefer using a telescope as it allows them to see farther.

Useful Equipment	What Is It Used for?
	With a pair of binoculars you will be able to see the craters on the Moon and the four largest moons of Jupit
	A compass will help you face the right direction when you are using the star charts.
	Use a flashlight to help you read t star chart. Putting red cellophane over the lightbulb end of the flashlight will prevent its light from affecting your night vision.

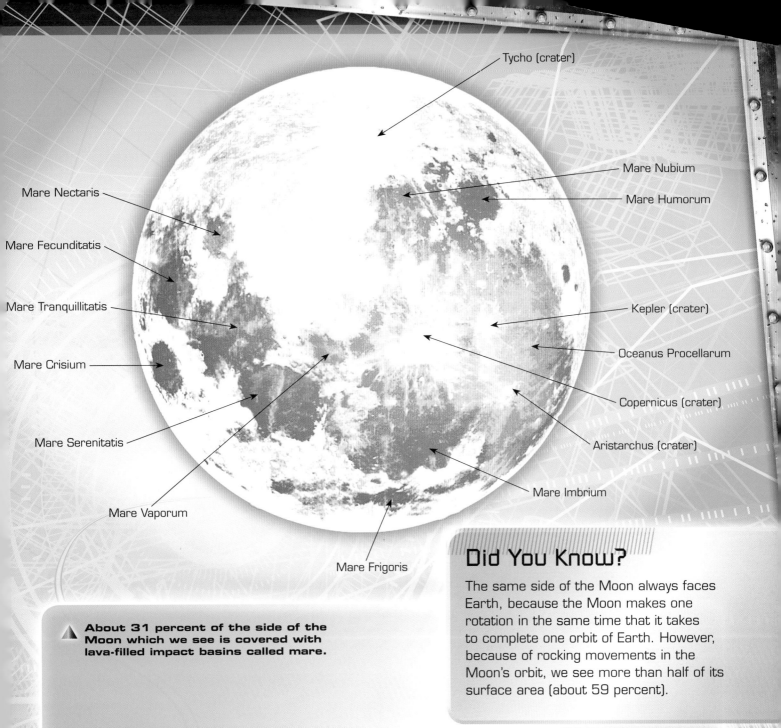

Tycho (crater)

Mare Nubium

Mare Humorum

Mare Nectaris

Mare Fecunditatis

Mare Tranquillitatis

Kepler (crater)

Oceanus Procellarum

Mare Crisium

Copernicus (crater)

Aristarchus (crater)

Mare Serenitatis

Mare Imbrium

Mare Vaporum

Mare Frigoris

⚠ About 31 percent of the side of the Moon which we see is covered with lava-filled impact basins called mare.

Did You Know?

The same side of the Moon always faces Earth, because the Moon makes one rotation in the same time that it takes to complete one orbit of Earth. However, because of rocking movements in the Moon's orbit, we see more than half of its surface area (about 59 percent).

Identifying Space Objects

With thousands of sources of light in the night sky, it can be difficult to identify space objects correctly. A bright light that is not marked on your star chart is probably a planet. A fast-moving streak of light could be a meteor, a piece of rock burning up as it travels through Earth's atmosphere. A **satellite**, such as the *Hubble Space Telescope* or the *International Space Station*, will appear bright and move steadily.

Moon Watching

The Moon is lit by reflected sunlight. It is the largest and brightest object in the night sky. The dark regions are called *mare*, which is Latin for sea. They are actually impact basins formed when objects crashed into the Moon. Dark lava eventually filled these basins. The lighter regions are called *terra*, meaning land. They are rugged mountain ranges.

CONSTELLATIONS

A constellation is an area of the sky that contains a group of stars that form a pattern when seen from Earth. Most groups of stars that make up constellations are extremely far apart in space. They just appear close when viewed from Earth.

The Origin of the Constellations

Many of the constellations we recognize today have their origins in ancient Greece. Ancient Greek astronomers named the patterns they saw in the stars after characters from Greek mythology, or legends. Today, the sky is divided into 88 official constellations whose boundaries are clearly defined. This means that every star belongs to just one constellation. Some well-known constellation star patterns are listed in the table below.

Constellation Star Pattern	Name(s)	Origin of Name	Visible From
	Crux (Southern Cross)	It is Latin for cross.	Southern Hemisphere
	Gemini	It is Latin for twins and associated with the twins Castor and Pollux from Greek mythology.	both hemispheres
	Orion	Orion was a hunter in Greek mythology.	both hemispheres
	Pegasus	Pegasus was a winged horse in Greek mythology.	both hemispheres
	Ursa Major (The Great Bear)	It is Latin for great bear and is associated with the Greek myth of Callisto, who was turned into a bear.	Northern Hemisphere

A group of stars that is not an officially recognized constellation is called an asterism. Well-known asterisms include the Plough (or Big Dipper) in Ursa Major, the Pleiades (or Seven Sisters) in Taurus, and the Saucepan in Pavo.

⚠ The Emu in the Sky is an Aboriginal Australian constellation which stretches across the Milky Way. The shape of the emu is formed by the dark dust clouds, not the stars.

The Zodiac

The path the Sun takes across the sky is called the ecliptic. The zodiac is made up of 12 of the constellations that are along the ecliptic path. It was one of the first celestial calendars, developed by the Babylonians during the first millennium BC. Some people believe that the sign of the zodiac under which a person is born affects his or her character.

Constellations around the World

In ancient China, astronomers divided the night sky into 31 regions, comprising 3 "enclosures" and 28 "mansions." In Indian astronomy, there are 27 *nakshatras*, or lunar mansions, which are divisions in the sky following the path of the Moon. Aboriginal Australian astronomers identified dark nebulae as animals. The most famous of these is Emu in the Sky.

THE NORTHERN SKY

The northern sky is the part of the sky visible from the Northern Hemisphere, which is north of the equator. Follow the instructions below to view constellations and stars in the northern sky.

How to Use the Star Chart

Find the current month at the edge of the star chart. Turn the book so that this month is at the bottom. Go outside at night and face south. You should be able to see most of the stars in the center and bottom half of the chart. At different times of the year, you will be able to spot different constellations and stars.

Spring (March to May)

Looking high in the sky to the south, the constellation of Leo, the lion, is easy to spot. Just look for the back-to-front question mark, which represents the lion's tail. Almost directly overhead is the constellation of Ursa Major, featuring the familiar pattern of stars known as the Plough or Big Dipper.

Summer (June to August)

Look low in the sky to the south to find the constellations of Sagittarius and Scorpius. Both constellations cut across the Milky Way, which is the hazy band of light that arcs across the entire sky. The Milky Way is brightest near Sagittarius because the center of the galaxy lies in this direction.

Did You Know?

The Northern Hemisphere is where most of Earth's land mass is found and the majority of its population. Light from towns and cities makes the stars more difficult to see. This is called light pollution.

Fall (September to November)

In fall, look for the constellation of Pegasus, the winged horse, high in the sky to the south. To the right of Pegasus, spot the tiny constellation of Delphinus, which looks like a dolphin leaping out of the water.

Winter (December to February)

Orion, the hunter, rides high in the sky in winter. Below and to the left of Orion is Sirius, the brightest star in the sky. It is close to the horizon at this time of year, and atmospheric effects make it sparkle particularly brilliantly.

THE SOUTHERN SKY

The southern sky is the part of the sky visible from the Southern Hemisphere, which is south of the equator. Follow the instructions below to view constellations and stars in the southern sky.

How to Use the Star Chart

Find the current month at the edge of the star chart. Turn the book so that this month is at the bottom. Go outside at night and face north. You should be able to see most of the stars in the center and bottom half of the chart. At different times of the year, you will be able to spot different constellations and stars.

Spring (September to November)

The winged horse, Pegasus, looms large in the springtime sky. To find it, look for four widely spaced stars in a rectangular shape. This is called the Square of Pegasus and represents the horse's chest. The stars on one side of this represent the front legs, and the stars on the other side represent the wings.

Summer (December to February)

Look north and find Orion, the hunter. The three bright stars in the center of Orion are called Orion's belt. Above the belt is a line of three stars known as Orion's sword. The middle star is actually not a star at all. It is the Orion **Nebula**, a massive cloud of dust and gas in which new stars are being formed.

Fall (March to May)

High in the sky is the constellation Crux (the Southern Cross). It points toward the south celestial pole, which is the point in the sky directly above the South Pole. Beside Crux is the large, bright constellation of Centaurus. Centaurus contains Alpha Centauri, a triple star system. One of its stars, Proxima Centauri, is the nearest star to the Sun.

Winter (June to August)

Look high in the sky to the north to find the constellations of Sagittarius and Scorpius. Both constellations cut across the Milky Way, the hazy band of light that arcs across the entire sky. The Milky Way is brightest near Sagittarius because the center of the galaxy lies in this direction.

Did You Know?

Fewer people live in the Southern Hemisphere than in the Northern Hemisphere. There is also less development, which means light pollution is less of a problem and conditions are better for stargazing.

THE BUILDING BLOCKS OF MATTER

Objects that take up space and have mass are called matter. All matter in the universe, including every star, planet, and moon, is made up of the same basic building blocks—atoms. Atoms join together to form elements and compounds.

Atoms

Atoms are made up of particles called protons, electrons, and neutrons. Protons have a positive **electric charge**, electrons have a negative electric charge, and neutrons have no electric charge at all. The protons and neutrons cluster together in the central part of the atom, called the nucleus, and the electrons "orbit" the nucleus.

▼ **Most elements were created inside supernova explosions such as this one, observed in 1994 by the *Hubble Space Telescope*. The supernova is on the bottom left and outshines the nearby galaxy.**

Did You Know?

Almost all of the elements that exist today were created inside dying stars billions of years ago.

Subatomic Particles

A subatomic particle is any particle that is smaller than an atom. The type of subatomic particle that is most important to scientists is the photon. It is produced by stars and is the basic unit of light and all other forms of electromagnetic radiation. The neutrino is another subatomic particle produced by stars. More than 50 trillion neutrinos from the Sun pass through the human body every second.

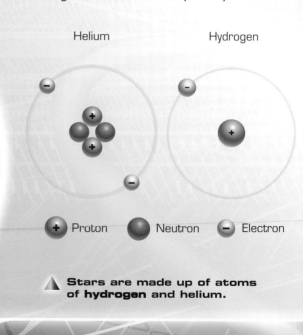

Helium   Hydrogen

+ Proton   Neutron   – Electron

▲ **Stars are made up of atoms of hydrogen and helium.**

18

This image by *SOHO (Solar and Heliospheric Observatory)* shows the Sun during a violent eruption known as a coronal mass ejection (CME). During a CME, enormous amounts of plasma blast into space at millions of miles per hour.

Water molecule

Oxygen atom

Hydrogen atom

One molecule of water is made up of one atom of oxygen and two atoms of hydrogen.

Elements and Compounds

An element is a substance that is made up entirely of one type of atom. An example of an element is hydrogen. Its atoms all have one proton and one electron.

A compound is a substance that is made up of more than one type of atom. In a compound, atoms of different types join together to form larger building blocks called molecules. An example of a compound is water. Its molecules are made up of two atoms of hydrogen and one of oxygen.

States of Matter

When matter gains or loses energy, it changes from one state to another. The most familiar states are solid, liquid, and gas. The fourth state is plasma. Plasma is superheated gas that has an electric charge. It is by far the most common state of matter in the universe. Plasma is what the Sun and most other stars are made of.

THE FUNDAMENTAL FORCES

A force is a push or a pull. There are four fundamental forces that are responsible for all interactions between matter: the weak nuclear force, the strong nuclear force, gravity, and electromagnetism.

Nuclear Forces

The weak nuclear force and the strong nuclear force are two very important forces that act within the nucleus of an atom. These are called nuclear forces.

The Weak Nuclear Force

The weak nuclear force acts within the nucleus of an atom, over a very short distance that is roughly the same as the diameter of an electron. The weak nuclear force is involved in the process of **nuclear fusion**, which makes stars, like the Sun, shine.

▼ Bright, blue, newly formed stars are seen here inside a **nebula** found in a galaxy called the Small Magellanic Cloud. The weak nuclear force helps nuclear fusion occur in the core of each star. This releases energy, which pushes the edges of the nebula, creating a dusty border.

The Strong Nuclear Force

The strong nuclear force holds neutrons and protons together in the nucleus of an atom.

Reactors in nuclear power stations have to overcome the strong nuclear force in order to split the atomic nucleus. This releases an enormous amount of energy.

The strong nuclear force is also involved in nuclear fusion.

The strong nuclear force is behind the vast amount of energy released by stars, such as these young ones in the constellation of Corona Australis. This energy is what makes the stars shine.

Did You Know?

The strong nuclear force is well named. It is 10 trillion times stronger than the weak nuclear force.

Gravity

All matter in the universe pulls on, or attracts, all other matter. This fundamental force is known as gravity.

How Gravity Works

Even though gravity is the weakest of the fundamental forces, it plays an important role in shaping the universe. This is partly because it affects all matter (unlike the other fundamental forces), and partly because it acts over such enormous distances. The greater the mass of an object, the greater the gravitational pull it exerts.

The small ball, which is closer than the medium-sized ball, is drawn toward the large ball. It will move closer and closer until it finds a point at which it will stop. In space, this is the point at which the smaller body goes into orbit around the larger body.

The large ball creates a dent in the gray area around it.

This ball is not as affected by the dent as the small ball because it is farther away.

Imagine the gray area is the universe. Heavy matter, represented by the large ball, pulls lighter matter, such as the smaller balls, toward it. This is how gravity works.

Did You Know?

The Sun's gravity is immense. A person
who weighs 110 lb (50 kg) on Earth would
weigh more than 2,866 lb (1,300 kg) on
the surface of the Sun!

The Effects of Gravity

Everyday objects exert gravitational pull on
each other but as these objects are not massive
enough, the force is not seen or felt. The
effects of gravity are most clearly seen at large
scales. Among other things, gravity controls
the orbit of the Moon around Earth, keeps
the planets in orbit around the Sun, and is
responsible for the formation of galaxies.

Gravity and Weight

The gravitational pull exerted by a planet or
other celestial body affects the weight of
objects on its surface. The greater the mass of
this body, the greater the gravitational force it
exerts and the greater the weight of objects on
its surface. The table, right, shows how much a
person who weighs 110 pounds (50 kg) on Earth
would weigh around the solar system.

Planet or Moon	A person who weighs 110 lb (50 kg) on Earth would weigh...
Jupiter	260.6 lb (118.2 kg)
Mars	41.4 lb (18.8 kg)
Mercury	41.7 lb (18.9 kg)
The Moon	18.3 lb (8.3 kg)
Neptune	123.9 lb (56.2 kg)
Saturn	117.3 lb (53.2 kg)
Uranus	97.9 lb (44.4 kg)
Venus	99.9 lb (45.3 kg)

Electromagnetism

Electricity and magnetism are two aspects of a single force known as electromagnetism. It is one of the four fundamental forces that shape the universe.

▼ **Charged particles from the Sun often interact with Earth's magnetic field near the poles. This produces a pattern of lights called an aurora. This aurora was photographed in Alaska.**

How Does Electromagnetism Work?

Electromagnetism acts between all objects that have an electric charge. A changing magnetic field creates an electric current, just as an electric current generates a magnetic field. As a star rotates on its axis, for example, an electric current is created. This in turn generates a magnetic field. The same thing happens if a planet's core is made of a substance that conducts electricity.

What Is Electromagnetic Radiation?

Electromagnetic radiation describes electromagnetic disturbances that travel in the form of waves. This radiation can be defined in terms of wavelength (how long each wave is) and frequency (how many waves there are in a given period of time). All types of electromagnetic radiation travel at the same speed, so longer wavelengths mean lower frequencies, and shorter wavelengths mean higher frequencies.

What Is the Electromagnetic Force?

The electromagnetic force is the force that an electromagnetic field exerts on electrically charged particles at an atomic and subatomic level. It is the electromagnetic force that holds electrons and protons together in atoms and holds atoms together to make molecules. It is also the force involved in chemical reactions.

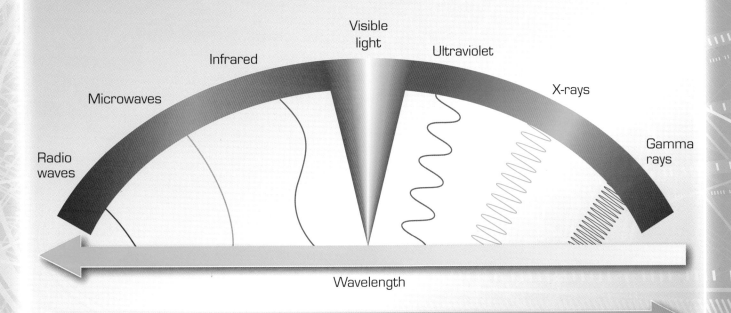

This diagram shows the various types of radiation that make up the electromagnetic spectrum. Radio waves have the longest wavelengths and lowest frequencies, and gamma rays have the shortest wavelengths and highest frequencies.

Did You Know?

Much of what scientists know about space has been discovered by observing sources of electromagnetic radiation.

DARK MATTER

Dark matter is matter that is not visible through telescopes, because it does not emit or reflect light. Scientists believe it may account for up to 90 percent of matter in the universe.

How Do We Know About Dark Matter?

If dark matter cannot be seen, how do scientists know it exists? Even though it is not visible directly, dark matter can be detected by its effect on light.

▼ The *Hubble Space Telescope* captured this image of a galaxy cluster with a ring of dark matter surrounding it.

Imagine light traveling from a distant galaxy toward Earth. If an extremely massive, dark object were to lie between the galaxy and Earth, light from the galaxy would bend. Stars and galaxies that have had their light bent in this way appear curved.

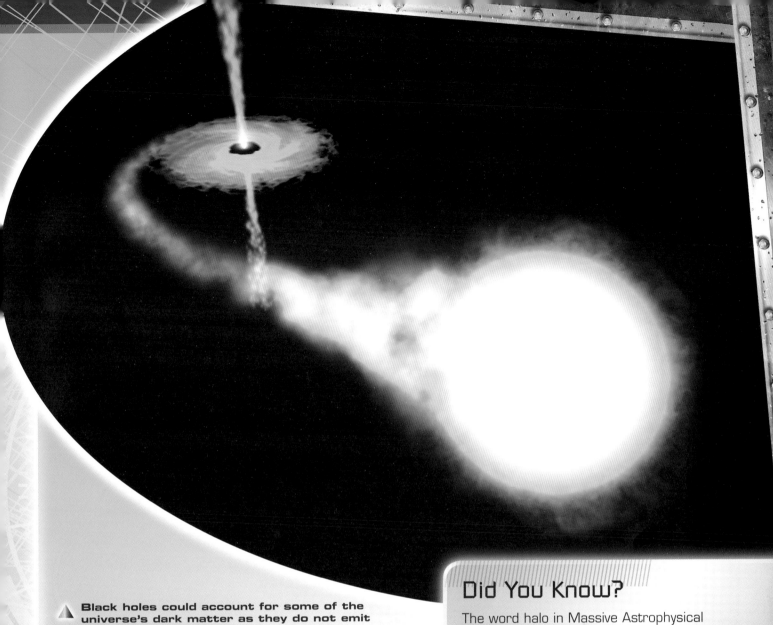

▲ Black holes could account for some of the
universe's dark matter as they do not emit
or reflect light. This artist's impression
shows a black hole (in the background)
pulling a star toward it.

Did You Know?

The word halo in Massive Astrophysical
Compact Halo Objects (MACHO) refers to
where most MACHOs are likely to be—in
the halos of galaxies. A halo is a large band
of dark matter that surrounds the visible
part of a galaxy.

MACHOs and WIMPs

Scientists have various theories about what
dark matter might be. One theory suggests
MACHOs (Massive Astrophysical Compact
Halo Objects). These are large, dark, extremely
dense objects such as burned-out stars and
black holes. Another theory suggests WIMPs
(Weakly Interacting Massive Particles). These
are hypothetical subatomic particles, similar to
neutrinos, which could have a tiny mass but
could exist in huge quantities.

Neutrinos

Neutrinos are a type of subatomic particle
produced by stars in vast numbers. Even
though they are common, they are very
difficult to detect. Scientists once thought
neutrinos had no mass at all, but experiments
have proved that they do have a very tiny
mass. This means neutrinos could account for
at least some dark matter.

SPACE AND TIME

In everyday life we think of space and time as separate things, but scientists view them as two parts of a single **phenomenon** called **space-time**.

Space-Time

By combining space and time into a single concept, scientists are better able to describe how the universe works on a small and large scale. Scientists define space-time as having four dimensions. These are the usual three dimensions of space (length, height, and width) and one dimension of time.

Time Dilation

Time dilation is a theory that the passage of time slows down as a moving object approaches the speed of light.

Time is dependent on certain factors. The rate at which one watch ticks compared to another depends on how fast the person wearing the watch is traveling in relation to the other. In everyday life, the difference is so tiny that it is hardly measurable. However, if one could travel close to the speed of light, time dilation would become significant.

1 Jane and Joe are twins.

5 When Jane arrives at home, after what seems to her like a short time, she finds that many years have passed on Earth, and Joe has become an old man.

2 Jane leaves Earth in a spacecraft while Joe stays at home.

3
Jane's spacecraft travels in a straight line at almost the speed of light.

4
Eventually, Jane slows down, turns around and heads back home.

⚠ This illustration shows how time dilation could become a problem for future space travelers.

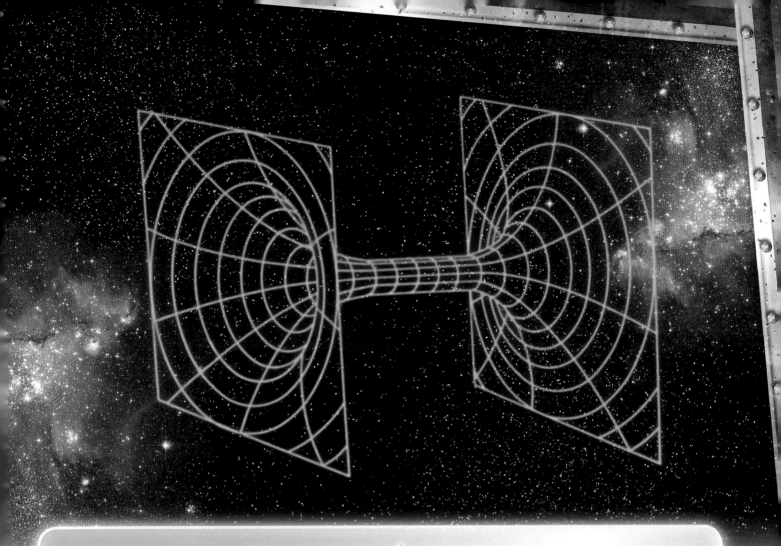

Time Travel

Some scientists have argued that if space is linked to time, perhaps it is possible to travel through time as well as through space. They have proposed several ways in which time travel might be possible. One of the most interesting ideas is the concept of wormholes.

Wormholes

Imagine space-time as a flat piece of paper. Fold this paper gently in half. Now, to travel the length of the paper from end to end might take some time, but if a wormhole existed between the fold, the distance between the two ends would be shortened considerably. By entering a wormhole, it might be possible to travel from one part of the universe to another, or from one universe to another. There is no evidence that wormholes exist, but they are theoretically possible.

This illustration shows how a wormhole could provide a bridge between one universe (in blue) and another (in red).

Problems with Time Travel

Time travel raises many paradoxes, or contradictions. Here is an example. A scientist invents a time machine. She uses it to travel back in time by one hour. When she realises that the time machine works, she panics and destroys it while in the past. However, with the machine destroyed, she is unable to travel back to the present and could not have traveled back in time in the first place! Such paradoxes have led many scientists to suggest that time travel, or at least traveling to the past, is impossible.

THE FUTURE OF SPACE SCIENCE

In the future, new technological developments will improve the quality of observations from Earth and in space. With better observational data, scientists may be able to develop better theories explaining how the universe works and prove current theories right or wrong.

Observational Space Science

One exciting area of current research is the search for **extraterrestrial** life. So far, more than 300 **exoplanets** have been discovered. Most of them are **gas giants**, but as detection techniques improve, scientists expect to find smaller, rocky planets too. A rocky planet located far enough from its parent star and with the right type of atmosphere could support extraterrestrial life.

Did You Know?

One way scientists search for extraterrestrial life is by scanning the sky with radio telescopes. They hope to receive radio signals from intelligent life forms.

Theoretical Space Science

At the moment, there are two scientific theories explaining the way the universe works. One theory (quantum mechanics) explains how subatomic particles behave, and the other theory (general relativity) explains how larger objects, such as stars and galaxies, behave. Some scientists are trying to come up with a single theory that will explain everything, but perhaps such an explanation does not exist.

Humans have long been fascinated by the possible existence of extraterrestrial life and have expressed this interest through television and the movies. In *ET the Extraterrestrial*, an alien being is left behind on Earth and befriends a young boy.

GLOSSARY

atmosphere
the layer of gases surrounding a planet, moon, or star

big bang theory
the theory that the universe expanded from an extremely dense and hot state and continues to expand today

black holes
regions of space where gravity is so powerful that nothing can escape, not even light

cosmic background radiation
a form of electromagnetic radiation that fills the universe but is not associated with any space object

dwarf planet
a small planet-like body that is not a satellite of another body but still shares its orbital space with other bodies

electric charge
a property some subatomic particles have, which can either be positive or negative

electromagnetic radiation
waves of energy created by electric and magnetic fields

electromagnetic spectrum
the range of all possible wavelengths of electromagnetic radiation, including radio waves, microwaves, infrared, visible light, ultraviolet, X-rays, and gamma rays

exoplanets
planets that are outside our solar system, orbiting a star other than the Sun

extraterrestrial
coming from outside Earth, or a being not from Earth

galaxy
a large system of stars, gas, and dust held together by gravity

gas giants
large planets made mostly of gas and with a metal or rock core, such as Jupiter, Saturn, Uranus, and Neptune

gravity
the strong force that pulls one object toward another

hydrogen
the lightest chemical element and the most common element in the universe

nebulae
clouds of gas or dust in space

neutron stars
very hot, small, and dense stars, left behind by supernovae

nuclear fusion
the process in which the nuclei of two or more atoms fuse together to form a single atom with a heavier nucleus, releasing huge amounts of energy

orbited
followed a curved path around a more massive object while held in place by gravity; the path taken by the orbiting object is its orbit

phenomena
something that can be observed, especially something unusual or interesting

pulsars
rotating neutron stars that emit a beam of electromagnetic radiation, which can only be seen when the beam is pointing towards Earth

quasars
the highly energetic cores of active galaxies, believed to be powered by supermassive black holes

satellite
a natural or artificial object in orbit around another body

solar system
the Sun and everything in orbit around it, including the planets

space-time
a concept that combines the three dimensions of space (length, height, and width) with the remaining dimension of time

supernovae
exploding stars

time dilation
a theory that the passage of time slows down as a moving object approaches the speed of light

INDEX